LIFE CYCLES IN REVIEW

BIRD

LIFE CYCLES

BY THERESA EMMINIZER

Enslow PUBLISHING

Please visit our website, www.enslow.com. For a free color catalog of all our high-quality books, call toll free 1-800-398-2504 or fax 1-877-980-4454.

Cataloging-in-Publication Data
Names: Emminizer, Theresa.
Title: Bird life cycles / Theresa Emminizer.
Description: Buffalo, NY : Enslow Publishing, 2025. | Series: Life cycles in review | Includes glossary and index.
Identifiers: ISBN 9781978539730 (pbk.) | ISBN 9781978539747 (library bound) | ISBN 9781978539754 (ebook)
Subjects: LCSH: Birds–Life cycles–Juvenile literature.
Classification: LCC QL676.2 E535 2025 | DDC 598.156 -dc23

Published in 2025 by
Enslow Publishing
2544 Clinton Street
Buffalo, NY 14224

Portions of this work were originally authored by Bray Jacobson and published as *A Look at Life Cycles: Bird Life Cycles*. All new material in this edition is authored by Theresa Emminizer.

Designer: Leslie Taylor
Editor: Theresa Emminizer

Photo credits: Cover (photo) Victor Tyakht/Shutterstock.com; series art (blue/green background) Denis Ka/Shutterstock.com; series art (grunge shapes) Midstream/Shutterstock.com; p. 5 duangnapa_b/Shutterstock.com; p. 7 (top) Milan Zygmunt/Shutterstock.com, (bottom) Wirestock Creators/Shutterstock.com; p. 9 Ryan Boedi/Shutterstock.com; p. 11 Birdiegal/Shutterstock.com; p. 13 Marten_House/Shutterstock.com; p. 15 Vladimir Turkenich/Shutterstock.com; p. 17 Donna A. Herrmann/Shutterstock.com; p. 19 WildMedia/Shutterstock.com; p. 21 Mike Laptev/Shutterstock.com; p. 23 Nicola_K_photos/Shutterstock.com; p. 25 Petr Simon/Shutterstock.com; p. 27 Chase D'animulls/Shutterstock.com; p. 29 John Klavitter/U. S. Fish and Wildlife Service/File:Laysan albatross fws.JPG/commons.wikimedia.org; p. 30 Mis vector/Shutterstock.com.

Printed in the United States of America

CPSIA compliance information: Batch #CSENS25: For further information, contact Enslow Publishing at 1-800-398-2504.

Contents

Words in the glossary appear in **bold** the first time they are used in the text.

Birds of a Feather

What makes a bird a bird? Birds are warm-blooded, vertebrate (backboned) animals. They have wings and beaks, which are hard mouthparts used to break apart food. Most importantly, they have feathers! Having feathers is what sets birds apart from all other animals.

LEARN MORE

WARM-BLOODED ANIMALS HAVE A BODY **TEMPERATURE** THAT STAYS ABOUT THE SAME NO MATTER THE WEATHER.

Many Species

There are thousands of different species, or kinds, of birds. They live in all sorts of different habitats, or natural homes, throughout the world. Each species has its own set of special **behaviors** that make it unique, or one-of-a-kind.

LEARN MORE

MOST BIRD SPECIES CAN FLY, BUT SOME CAN'T! OSTRICHES, KIWIS, EMUS, AND PENGUINS ARE ALL FLIGHTLESS BIRDS. PENGUINS ARE VERY GOOD SWIMMERS. EMUS ARE FAST RUNNERS.

ostrich (largest bird)

hummingbird (smallest bird)

Finding a Mate

Mates are two animals that come together to make babies. Male birds try to attract, or draw in, female birds to be their mate by showing off their brightly colored feathers. They may also bring females food or do a little dance!

LEARN MORE

MALE BIRDS OF PARADISE SING AND PUT ON AN EYE-CATCHING SHOW TO ATTRACT MATES. MALE PEACOCKS FAN, OR SPREAD OUT, THEIR LONG, BRIGHT TAIL FEATHERS AS THEY DANCE.

bird of paradise

Building a Nest

Each bird species builds its own special kind of nest. Eagles build big nests called aeries high on cliffs or mountains. Puffins, kingfishers, and bank swallows make **burrow** nests in the ground. Woodpeckers make nests in cavities, or holes, in tree trunks.

LEARN MORE

IN MANY SPECIES, FEMALES BUILD NESTS, BUT SOME KINDS OF MALE BIRDS HELP THEIR MATES. SOME BIRDS USE GRASS AND MUD. BIRDS MAY ALSO GATHER PAPER OR OTHER HUMAN-MADE MATTER.

eagle nest

Laying Eggs

Mother birds lay eggs. A group of eggs is called a clutch. Some birds lay only one or two eggs, others lay as many as 50! Mother birds then need to incubate the eggs, which means keeping them warm until they **hatch**.

LEARN MORE

SOME BIRDS WAIT UNTIL ALL THE EGGS ARE LAID BEFORE INCUBATING. OTHERS BEGIN INCUBATING THE FIRST EGG RIGHT AWAY. THAT MEANS BABY BIRDS IN THE SAME CLUTCH MAY BE BORN ON DIFFERENT DAYS!

Brood Parasites

Some bird species do something different. They lay their eggs in other birds' nests! They may even lay more than one egg in more than one nest. These birds are called brood parasites. They then leave their young to be cared for by other mothers!

LEARN MORE

THE COMMON CUCKOO IS A SPECIES OF BROOD PARASITE. THE BABY CUCKOO IS OFTEN BIGGER THAN THE MOTHER BIRD'S OWN BABIES, MEANING IT EATS MORE FOOD AND TAKES UP MORE SPACE.

baby cuckoo

Hatchlings

After incubation, the baby birds hatch. Baby birds are called hatchlings. Many hatchlings, including baby songbirds and seabirds, are born blind, featherless, and helpless. They need their parents to feed and care for them.

LEARN MORE

HATCHLINGS ARE ALSO CALLED CHICKS. CHICKS MAKE A LOT OF NOISE, WHICH CAN DRAW PREDATORS TO THE NEST. PREDATORS ARE ANIMALS THAT EAT OTHER ANIMALS.

Nestlings

In many species, both mother and father birds feed their young. Once hatchlings can open their eyes, they're called nestlings. Some parents feed nestlings through regurgitation. Regurgitation is when a parent bird eats food, then spits it back up to feed its young.

LEARN MORE

SOME SPECIES, SUCH AS SONGBIRDS, FEED AND CARE FOR THEIR YOUNG FOR JUST A FEW WEEKS. OTHERS, SUCH AS THE ALBATROSS, CARE FOR THEIR YOUNG FOR MANY MONTHS!

Fledglings

Fledging is leaving the nest. Once a nestling has grown flight feathers, it's called a fledgling. Fledglings are usually still fed and cared for by their parents. Mother birds may nudge fledglings, or give them a soft push, out of the nest.

LEARN MORE

IF YOU SEE A YOUNG BIRD ON THE GROUND, IT MIGHT BE A FLEDGLING JUST LEARNING TO FLY! SOMETIMES PEOPLE MISTAKE FLEDGLINGS FOR HURT BIRDS. DON'T WORRY! THEY MAY JUST BE RESTING.

Learning to Fly

Birds don't hatch knowing how to fly. They have to learn how to do it! Parents usually stay nearby to help. Fledglings fall, flap their wings, and try over and over again. It takes time to learn, but birds have strong **instincts**.

LEARN MORE

TAKING OFF AND LANDING ARE JUST TWO PARTS OF FLYING. FLEDGLINGS ALSO HAVE TO LEARN HOW THE WIND AFFECTS FLIGHT. BIRDS USE THEIR TAIL FEATHERS TO **STEER**.

Migratory Birds

Migrating is moving from one area to another for feeding or having babies. There are 4,000 known species of migratory birds. The Arctic tern makes the longest migratory journey, or trip. It flies more than 49,700 miles (79,984 km) each year.

LEARN MORE

ARCTIC TERNS MAY LIVE LONGER THAN 30 YEARS. ADDED UP OVER THAT LIFETIME, THEIR MIGRATORY FLIGHTS CAN EQUAL THE NUMBER OF MILES IT WOULD TAKE TO GO TO THE MOON AND BACK THREE TIMES!

Arctic tern

Special Changes

Some migratory birds can change their bodies in special ways. Ruby-throated hummingbirds double their body weight to get ready for their 2,000 mile (3,219 km) migration. This **fuels** their heart, which beats 1,000 times per minute, and wings, which beat 3,000 times per minute.

LEARN MORE

SOME BIRDS CHANGE THEIR SLEEP TO HELP THEM TRAVEL. SWAINSON'S THRUSHES GO INTO A SLEEP-LIKE STATE SO THEY CAN REST WHILE FLYING! HALF THEIR BRAIN STAYS AWAKE TO KEEP THEM SAFE.

ruby-throated hummingbird

Unlike Any Other

It's believed that birds first showed up on Earth about 160 million years ago. Over time, they **evolved** into the thousands of species we know today. Each bird's journey from egg to adulthood is **fascinating**!

LEARN MORE

THE OLDEST-KNOWN WILD BIRD IS AN ALBATROSS NAMED WISDOM. WISDOM IS OVER 70 YEARS OLD. SHE MIGRATES TO THE SAME NEST SITE EVERY YEAR. WISDOM IS BELIEVED TO HAVE FLEDGED UP TO ABOUT 30 CHICKS.

Wisdom, the albatross

Life Cycle of the American Robin

Adult American robins find mates.

They build nests.

Mother robins lay eggs.

They incubate their eggs.

Baby birds hatch from the eggs.

Hatchlings grow into nestlings.

Hatchlings grow flight feathers and become fledglings.

Fledglings learn to fly and leave the nest.

Some American robins may migrate to mate, nest, lay eggs, or find food.

Glossary

behavior: The way an animal acts.

burrow: A hole in the ground.

evolve: To change over time to be better adapted to surroundings.

fascinating: Very interesting.

fuel: To power, or use food to make energy.

hatch: To come out of an egg.

instinct: An inborn reaction or behavior that aids in survival.

steer: To direct or move something a certain way.

temperature: How hot or cold something is.

For More Information

BOOKS

Bradley, Doug. *Animal Migration*. Buffalo, NY: Britannica Educational Publishing, 2024.

Hughes, Sloane. *20 Things You Didn't Know About Bird Adaptations*. New York, NY: PowerKids Press, 2023.

WEBSITE

Audubon for Kids
www.audubon.org/get-outside/activities/audubon-for-kids
Read articles, watch videos, and try fun activities to learn more about birds.

Publisher's note to educators and parents: Our editors have carefully reviewed this website to ensure it is suitable for students. Many websites change frequently, however, and we cannot guarantee that a site's future contents will continue to meet our high standards of quality and educational value. Be advised that students should be closely supervised whenever they access the internet.

Index